Science
Grades 4 to 8

THINKING ABOUT SCIE

Ecology

Written by Rebecca Stark

ISBN 9781-56644-062-2

Educational Books 'n' Bingo

Previously published by Educational Impressions, Inc.

Printed in the U.S.A.

Table of Contents

To the Teacher

Ecology is a comprehensive, fun-filled unit on ecology. Students learn important facts about biomes, ecosystems, food chains, and more. They are also presented with opportunities to practice crucial critical- and creative-thinking skills. A variety of types of activities are included: creative writing, research, analyzing, evaluating, and more.

A fun What's the Question? game is provided to reinforce the concepts learned in the unit. In addition, a crossword puzzle is included. The puzzle may be used to evaluate knowledge or just for fun!

Ecology: What Is It?

The word "ecology" comes from the Greek word *oikos*, meaning "home" or "place to live." Ecology is the study of organisms—plants and animals—and their environments. Their environments include other plants and animals and the non-living, physical surroundings. All of these interrelationships form ecological communities called ecosystems.

See how much you already know about ecology. Match the terms on the left with the definitions on the right.

_____ 1. Biome	A.	The part of Earth and its atmosphere where life exists.
_____ 2. Biosphere	B.	Arid, usually hot region getting less than about 10 inches of rain per year.
_____ 3. Carnivore	C.	A living community and its physical surroundings.
_____ 4. Coniferous	D.	A community of living organisms of a single major ecological region; named for dominant plant life.
_____ 5. Deciduous	E.	A plant-eating animal.
_____ 6. Desert	F.	Process by which green plants use the sun's energy to make food.
_____ 7. Ecologist	G.	A meat-eating animal; a predator.
_____ 8. Ecosystem	H.	An organism's place and function in a community.
_____ 9. Food Chain	I.	Referring to a tree that bears cones.
_____ 10. Herbivore	J.	Region that receives at least 100 inches of rain per year.
_____ 11. Niche	K.	A succession of organisms feeding on each other and passing on energy.
_____ 12. Photosynthesis	L.	A treeless area between the ice cap and the tree line of Arctic regions.
_____ 13. Rain Forest	M.	A scientist who studies organisms and their physical surroundings.
_____ 14. Tundra	N.	Referring to a tree that sheds all of its leaves seasonally.

Biomes

The prefix "bio-" comes from the Greek word *bios,* meaning "life." We refer to the thin layer extending about 5 to 6 miles (8 to 10 kilometers) above Earth's surface to a few feet (meters) below as the **biosphere.** It is that part of Earth's surface that is capable of supporting and sustaining life.

The biosphere is divided into large units called **biomes.** Generally speaking, a biome is a community of living organisms of a single major ecological region. Major biomes are **Northern Coniferous Forests, Temperate Deciduous Forests, Chaparrals, Tropical Rain Forests, Deserts, Tundra,** and **Grasslands.** The main factor distinguishing one biome from another is climate. Each biome has a particular pattern of rainfall, altitude, temperature range, seasonal changes, and changes in the length of daytime. These factors determine the dominant vegetation, which in turn determines the animal life that is found in the biome. At the borders are **ecotones**, where the plant and animal life of the adjoining biomes mix.

Activities

Create a "Bio-" Dictionary. List at least six words that contain the affix (usually a prefix) "bio-." Include the word "life" or "living" in each definition.

Define the following terms: coniferous, deciduous, chaparral, desert, grassland, rain forest, and tundra.

Explain why similar biomes are often found at similar latitudes.

Biomes are made up of smaller units called habitats. A habitat is the type of place a plant or animal species lives. For example, a whale's habitat is the sea. A cow's habitat is a pasture. Create a matching game of plants and animals and their habitats.

You have been asked by the Big City Zoo to design a suitable habitat for the animal of your choice. Price is no object. Research that animal and then make a detailed sketch of your ideas for a perfect habitat.

All Kinds of Organisms

We classify organisms into three different categories according to their function in the environment: producer, consumer, or decomposer. Plants are the only **producers.** Green plants manufacture their own food through a process called **photosynthesis.** Animals that eat those plants are called **primary consumers.** Those animals that eat the animals that eat the plants are called **secondary consumers.** If they in turn are consumed, the animals that eat them are called **tertiary consumers.** Bacteria and fungi are **decomposers.** They feed on dead organisms and break them down.

Activities

Analyze the following statement: "All the energy for life comes from the sun."

Draw a picture that illustrates the meaning of *predator* and *prey.* Label each.

Find out the meaning of *autotroph.* Find a synonym for this word in the introductory paragraph.

Define *herbivore, carnivore, omnivore, scavenger,* and *parasite.* Which are most humans?

Draw a poster in which you illustrate the terms you defined. Do not use humans for your poster.

Create three analogies using at least one of the terms printed in bold on this page in each analogy. An example follows:

Herbivore : Carnivore : : Zebra : Lion

What's Your Niche?

In any community there is competition for resources. Two or more species cannot exploit the same resource at the same time, in the same way, and in the same place. Each species, therefore, must have its own niche—its own function and place in the community.

THE MANGROVE COAST OF SOUTHERN FLORIDA

Many different species of heron inhabit the Mangrove Coast of southern Florida. At times they all can be seen feeding on the same shoal. (A shoal is an elevation of land coming close to but not above the surface of the water.) The reason they can co-exist is that each has its own niche. In addition to foraging in different parts of the shoal, each has a different means of capturing its prey.

The **green heron** waits motionless at the edge of the tidepool for an unknowing fish to pass.

The **common egret** stalks its prey. It walks slowly with its head extended and bill pointing downward, ready to grab its prey.

The **great blue heron** startles its prey by flicking its wings as it walks. The startled prey reacts and becomes more visible.

The **Louisiana heron** whirls, pivots, or races through the water before stopping suddenly, agitating its prey.

The **reddish egret** stirs up the water and then extends its wings, shading the water. The prey settles under the shade of the canopy formed by the wings with a false sense of security.

Activities

Different species are active at different times of the day. This allows more species to occupy the same space without too much interference or competition. Define diurnal, nocturnal, and crepuscular.

Judge the following statement: "The best adaptation is the ability to adapt."

Animals evolve in ways that will help them adapt to their particular niches. Create a poster that shows how birds' feet help them adapt to their niches.

Some animals have niches that are so specialized that their activities are extremely restricted. Explain why the panda and the koala have very restrictive niches.

Chains, Webs, and Pyramids

The dependence of living things upon other living things is most apparent in food chains. All living things need energy in order to survive. This energy gets passed along in food chains. A food chain is a kind of cycle. The energy and nutrients keep moving from non-living things to living things and back to non-living things.

Food chains—the series of steps of eating and being eaten—all begin with green plants. Plants change solar energy—the energy from the sun—to the chemical energy that is stored in food. They do this through a process called **photosynthesis**. Green plants use the energy from the sun to turn carbon dioxide and water—the non-living things they take from the environment—into sugar. Plants are called the producers because they are the only organisms capable of producing their own food. They produce the energy that will be passed on to the next link.

Animals are the consumers. When an animal eats a plant, nutrients and some of the energy that had been stored in the plant are passed on to the animal. If that animal is eaten by another, the nutrients and some of the energy that was passed on to it are then passed on to the second animal. Nutrients are the minerals and other elements that plants and animals need in order to develop properly. This passing of nutrients and energy from one living thing to another is what we call a food chain. With each link, some of the energy is lost. This is because energy, although it cannot be created or destroyed, can be changed. When stored energy is changed to working energy, some of it becomes heat energy. The heat energy goes into the air or water instead of passing on to the next link.

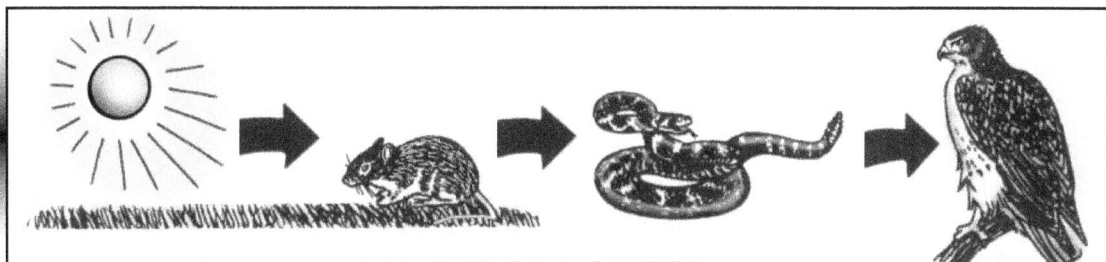

Sooner or later all living things die. When they do, they become part of a different kind of food chain. The dead organisms become food for bacteria and other decomposers. The decomposed, or decayed, plant and animal matter is broken down. Minerals and other nutrients from the decayed matter are released into the soil. Eventually, these nutrients are dissolved by water (a non-living thing) and are taken in by living plants. And so, the cycle continues.

MAIN TYPES OF FOOD CHAINS

1. The Predator Chain
In this type of food chain plant producers are eaten by plant-eaters, who are then eaten by larger carnivores.

2. The Parasite Chain
This chain includes large animals and the smaller animals which parasitize them and thus rob their energy store.

3. The Saprophyte Chain
This chain is primarily in the soil. Dead plants and animals are broken down by microorganisms.

FOOD WEBS: In any community there are a number of interrelated food chains. We call this complex of chains a food web.

Salt Marsh Community Food Web

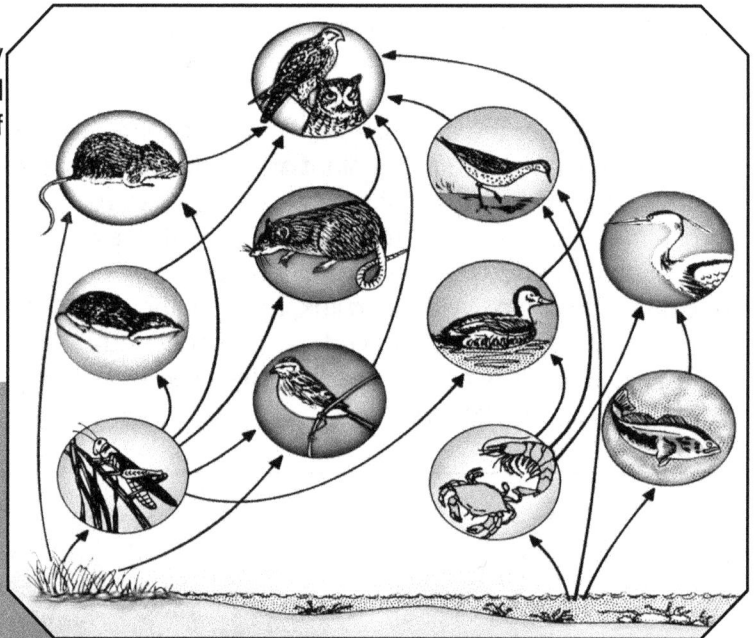

FOOD PYRAMIDS: Sometimes ecologists use pyramids to demonstrate the flow of energy from one energy level to the next. The base is made up of the producers, the green plants. Above the base would be the plant-eaters, the primary consumers. The third from the bottom would be those animals that eat the plant-eaters, the secondary consumers. The fourth, and often the top layer, would comprise the tertiary consumers, usually the top carnivores in the pyramid. Because energy is lost at each level, the top level can only support a few carnivores. The chart at the right is an example of a pyramid of numbers.

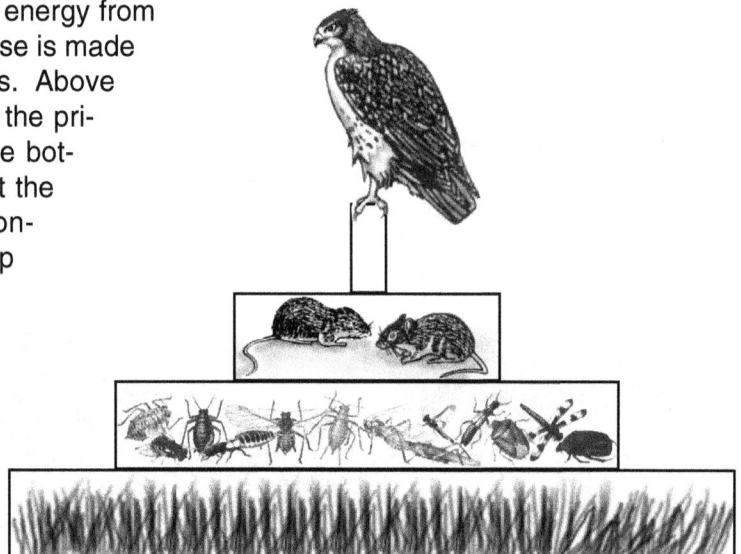

Activities

Most food chains have only three or four links. Rarely are there more than five. Draw a diagram that illustrates why this is so.

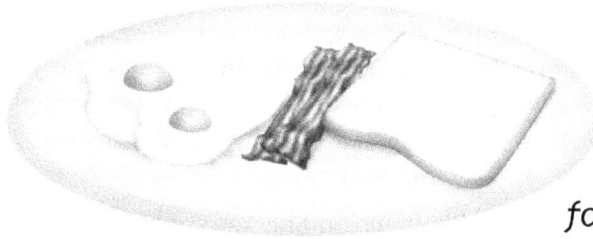

Draw a chart that shows when you are part of a 3-link food chain.

Draw a chart that shows when you are part of a 2-link food chain.

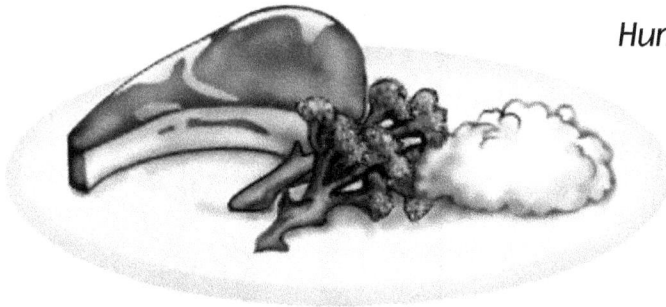

Humans are involved in many different trophic, or nutrition, levels. Make a list of everything you ate yesterday. At what level of the food chain were you for each. Were you involved in any parasitic relationships?

Draw a food web that might occur in a woodland environment.

A pyramid of numbers for a tree would not have a pure pyramid shape. Draw a sketch and explain.

Partnerships, Good and Bad

There are different types of relationships among living organisms. Some are beneficial to all those involved. Others are beneficial to one or more members. Still others are harmful to one or more members. When two different kinds of life-forms live together in close association, we call it **symbiosis.** Mutualism, commensalism, and parasitism are all forms of symbiotic relationships.

MUTUALISM

In mutualism, both life-forms benefit and neither is harmed. A lichen is an example of mutualism. It is made up of an alga and a fungus that grow together on a solid surface, such as a rock or a tree. NOTE: Sometimes "symbiosis" is used as a synonym for "mutualism."

COMMENSALISM

Commensalism is a relationship in which only one of the partners benefits. Here, too, neither is harmed. A barnacle that attaches itself to the body of a whale has a commensal relation with the whale. Only the barnacle benefits, but neither is harmed.

PARASITISM

If one organism in a relationship benefits from the association while the other is harmed by it, we call it a parasitic relationship. Lice living on a bird is an example of a parasitic relationship. The lice benefit; the bird is harmed.

Read the following descriptions. Decide which of the above relationships is being described.

1. Birds called oxpeckers pick the ticks off African buffaloes. The buffaloes benefit because the ticks that suck their blood are removed. Also, the birds make a lot of noise at any sign of danger, warning the buffaloes. The birds receive food, hair for their nests, and transportation.

_____

2. Some ants "farm" aphids. The ants stroke the body of the aphids, causing them to release honeydew, which they then carry back to feed to their young. The ants guard the aphids and protect them from predators. Some ants even round up their aphids and escort them to the safety of their nests.

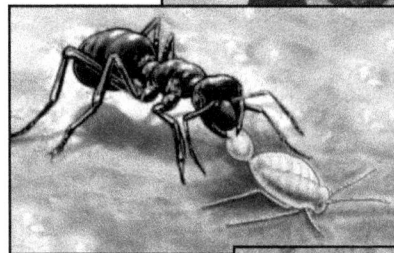

_____

3. Sea "lice" feed on a damsel fish.

_____

Ecosystems

Many scientists study ecology in terms of **ecosystems**. Ecosystems comprise both living organisms and the non-living things in their environment that sustain them, such as air, water, soil, and sunlight. Interactions in an ecosystem that are studied by ecologists include mineral cycling, energy flow, and population control.

MAJOR LAND ECOSYSTEMS

Tundra
Northern Coniferous Forest
Deciduous Woodland
Tropical Rain Forest
Grassland: Savanna
Grassland: Steppe
Desert

MAJOR AQUATIC ECOSYSTEMS

Lake and Pond
River and Stream
Open Ocean
Seashore
Estuary

Define steppe and savanna.

Define estuary.

How is the flow of energy in an ecosystem accomplished?

What is the main difference between a biome and an ecosystem?

The Arctic Tundra

The Arctic tundra is the treeless area between the ice cap and the tree line of the Arctic regions. It has a permanently frozen subsoil. Winters in the tundra are long and cold. Summers are short and cool. Most of the tundra receives less than fourteen inches of rain per year.

Activities

The average annual temperatures are below 32° F. Even in July the average is only about 39° F. Few days reach 50° F. Trees need at least 30 days a year when the daily average temperature is above 50° F. Write a syllogism based on this information.

A syllogism is a form of deductive reasoning consisting of a major premise, a minor premise, and a conclusion; for example, "Only birds have feathers. Robins have feathers; therefore, robins are birds."

For a very brief time, the tundra is very productive. Swans, geese, and other birds spend the summer there. Almost all are migratory. As the days grow short and the food supply begins to dwindle, they leave. The arctic tern has the longest migration route of any known animal species. It also gets more sunlight hours than any other species. Draw a map of its migration route.

North American caribou spend the summer in the Canadian tundra. Large herds travel over 1,000 miles, swimming across wide rivers and lakes to reach their destination. Reindeer have similar migrations in Siberia. Create a riddle about an animal that remains in the tundra year round.

High up in the mountains, conditions are in many ways similar to those of the Arctic. Usually, Earth's atmosphere acts like a blanket, preventing heat from escaping. Why is this blanketing effect reduced in the mountains? What other problems result for the same reason?

Northern Coniferous Forests

Coniferous forests are found throughout the temperate regions of the northern hemisphere. Conifers, also called softwoods, have needlelike leaves. Usually, they bear their seeds in cones. Most conifers, such as pine, spruce, and hemlock, are evergreens. Although they do shed their leaves, they loose them a little at a time. As some fall off, new ones grow to take their place; therefore, the trees always appear green.

Activities

The red crossbill can live only in the coniferous forest. Find out why.

Analyze the reasons that the favorable growing season is so short in the northern coniferous forest.

Giant redwoods found in coastal and northern California can live for 3,000 years. Create a haiku about a redwood.

A haiku is an unrhymed lyric poem having a fixed three-line form. The lines have 5, 7, and 5 syllables respectively.

Draw a picture of a coniferous tree. Explain how it is adapted to do well in spite of the short favorable growing system.

Explain what is meant by the taiga.

Temperate Deciduous Forests

Temperate deciduous forests are found in regions where there are moderate amounts of rainfall (30 to 80 inches per year), moderate temperatures, and cold winters. All four seasons are about the same length. "Deciduous" means that the trees shed their leaves at the end of the growing season. The term comes from a Latin word meaning "to fall off." Most deciduous trees are broadleaved. "Broadleaved" (also, broadleafed) means just what you would think: "having relatively broad leaves." Also called hardwoods, maples, oaks, hickories, and beeches are examples of deciduous trees.

Activities

Like in all communities, the inhabitants of the deciduous forest interact in many ways. One way is competition. Describe how trees compete with one another.

Describe the ways in which trees contribute to their environment.

PIGMENTATION

The yellow and orange pigments of autumn leaves are actually present even when the leaves are green. During the autumn, trees grow a layer of cells between the twigs and the leaf stems. This layer, called the abscission layer, prevents the leaves from receiving water and nutrients. The leaves stop producing chlorophyll. With no green chlorophyll to hide the yellow and orange pigments, the leaves appear yellow and orange. In some trees some of the sugars are made into reddish pigments, adding to the colorful array. When the abscission layer in a leaf dehydrates, the leaf falls off.

EXPERIMENT: (To be done with proper supervision)

1. Cut a **paper coffee filter** into a 1¹/₂" x 4" strip.

2. Place a **leaf** on the paper and rub the leaf in the center with a **pencil,** making a spot in the center of the strip. (Be sure it is in the center.)

3. With a **paper clip,** hang the filter paper from the pencil, which you have placed across the lip of a **jar.** Mark the jar to show where the bottom of the paper hangs.

4. Remove the paper and pour **rubbing alcohol or nail polish remover** so that it will just touch the bottom of the paper. Return the paper to the jar. (The alcohol should touch the paper but not the green dot.)

5. After about 30 minutes, the alcohol should travel upward. As it does, the green pigment should also travel upward. The yellowish streaks continuing past it should indicate the presence of another pigment.

Layers of the Forest

The forest is almost like an apartment building. The top story, called the **canopy,** comprises the leaves of the tallest trees, such as tall maples, oaks, and hickories. Below the canopy is the **understory.** The understory is made up of young trees and also mature trees of species that don't grow as tall, such as dogwoods. Just below the understory is the **shrub layer.** Shrubs are bushy plants with many woody stems. The **herb layer** is the first layer above the forest floor. Herbs are green plants with soft stems. Mosses, ferns, and mushrooms also live in this layer. Of course, beneath it all is the **forest floor.** In most deciduous forests the floor is covered with fallen leaves. It is in the bottom two layers that the widest variety of life can be found.

Activities

Explain the following statement: The understory of today is the canopy of tomorrow.

The climate is different at each level of the forest. Explain how trees affect the sunlight, temperature, wind, and rain amounts at each level.

Create a poster that illustrates the changes that take place in a deciduous forest as the cold winter approaches.

Draw a picture that represents a 3-link food chain that might take place in the canopy of a deciduous forest.

"Detritus" means "bits and pieces." In a forest it refers to the bits and pieces of dead plants and animals found on the forest floor. Draw a chart that explains how detritus is turned into a substance called "humus."

Tropical Rain Forests

Tropical rain forests occur in areas where rainfall amounts exceed 160 inches per year. Some even receive about 400 inches! It is hot all year round and there is no dry season in a tropical rain forest. The heated air rises and condenses, forming clouds and then precipitation. This never-ending supply of light, heat, and rain provide perfect conditions for rain forests to flourish.

There are various distinct layers in the tropical rain forest, which is dominated by broadleaved evergreen trees. The tops of these trees form a closed **canopy** over the forest. The height of the canopy is generally about 100 to 165 feet. A few trees rise about the canopy. These trees, called the **emergent level** because they "emerge from" the canopy, may grow to about 250 feet! Below the canopy level are the shorter trees of the **understory.** Because little sunlight gets through the thick canopy, only those species of trees which can grow in the shade are able to survive in the understory. The lack of sun also causes the **undergrowth,** or the bottom 15 feet, to be rather sparse. Patches of plants grow in the places where the sun manages to penetrate. At the very bottom, of course, is the **forest floor.**

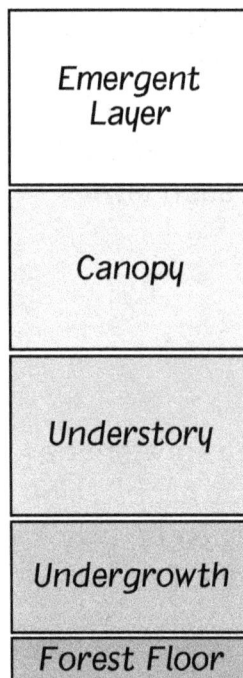

| Emergent Layer |
| Canopy |
| Understory |
| Undergrowth |
| Forest Floor |

Rain forests are very important to us. Although only about six percent of the biosphere is covered by rain forests, they are home to more than half of the known plant and animal species! Some experts believe they are home to as many as 90 percent of the world's species! Many species are still to be identified!

Strange as it may seem, tropical soil is extremely poor. That's because the heavy rains wash away the nutrients. In spite of this, however, tropical vegetation flourishes. It is helpful that tropical plants have very shallow roots. Shallow roots absorb the nutrients more quickly than roots that have to wait for the nutrients to seep down to lower levels.

The reason such poor soil can produce such lush vegetation is that the tropical rain forest is a closed, self-sustaining system. Explain.

Activities

Rain forests of the Old World tropics contain larger plants and animals than those of the New World tropics. Explain what is meant by "Old World" and "New World" tropics.

The Amazon Rain Forest is the largest in the world. It is spread over nine countries. Name the countries.

Find out why pharmaceutical companies might be interested in rain forests. Create a poster that one of these companies might distribute to encourage people to Save the World's Rain Forests.

Epiphytes are very common in the tropical rain forest. They are sometimes called "hitchhikers." Explain why and cite examples.

Referring to epiphytes as hitchhikers is a metaphor. Create an original metaphor about an animal or plant of the rain forest or about the rain forest itself.

Cite an example of a large mammal whose natural habitat is the Old World tropics.

There are more arboreal animals in the New World tropics. Define "arboreal." Give three examples of arboreal animals of the New World tropics.

Lianas and buttresses are unique to the rain forest. Explain what is meant by each term. Draw a split picture illustrating each.

Some people mistakingly call rain forests jungles. The floor of a jungle is covered with tangled undergrowth. The floor of a true rain forest has hardly any undergrowth at all. Guess why.

It is clear that animals of the rain forest are dependent upon the plants. Draw a poster that illustrates why the plant life depends upon the animals.

Rain Forest Fact or Opinion?

People make many statements regarding the rain forest. Some are based on fact and some are just opinion. Decide whether each of the following statements is fact or opinion. Write an F or an O in the space in front of each statement.

_____ 1. Poor, developing nations have a right to exploit the resources of the rain forest without regard to the amount of destruction.

_____ 2. Most of the world's tropical rain forests are found in a broken band around the equator.

_____ 3. The largest rain forest area is centered around the Amazon River in South America.

_____ 4. Poachers should be severely punished if caught.

_____ 5. The rain forest is a closed, self-sustaining system.

_____ 6. Rain forests should be preserved because of their wondrous beauty.

_____ 7. Lumber is the largest export of the rain forest.

_____ 8. Products from the rain forest include dyes, oils, shellac, and disinfectants.

_____ 9. Wicker furniture, which is made from lianas from the rain forest, is very attractive.

_____ 10. It is important to protect endangered species, such as the orangutan and the jaguar, from becoming extinct.

_____ 11. Many rain forests are rich in minerals, such as aluminum, tin, copper, iron, and uranium.

_____ 12. *Garimpeiros,* Brazilian prospectors who mine for gold in the rain forest illegally, should be better controlled by the government.

_____ 13. Rain forests are more beautiful than temperate deciduous forests.

_____ 14. Tropical rain forests are hot year round and have no dry season.

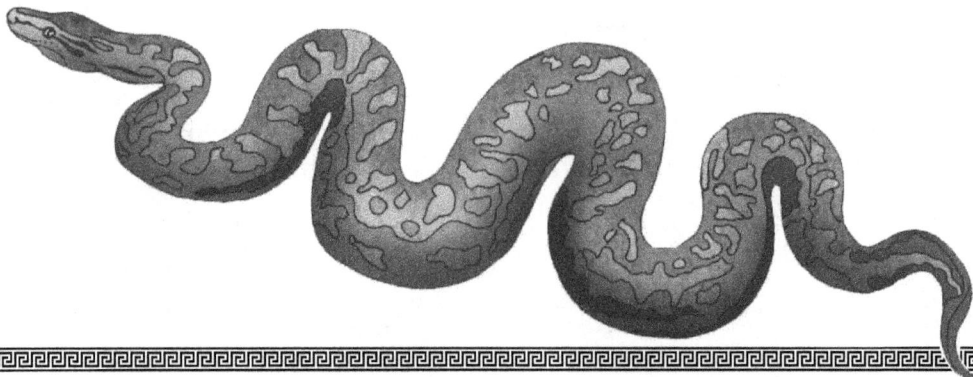

Foods from the Rain Forest

The following foods originated in the rain forest. Put them into groups. Stretch your imagination to think of some unusual groupings. You must have at least two items in a group.

coffee
chocolate
oranges
papayas
cloves
cashews

grapefruit
pineapples
ginger
black pepper
bananas
allspice

vanilla
cinnamon
nutmeg
paprika
eggplant
avocado

mangoes
tomatoes
corn
peanuts
sugar cane
limes

Animals of the Rain Forest

Choose from the names in the box to complete these sentences about animals of the rain forest. Use your dictionary or encyclopedia for help if necessary.

anaconda	armadillos	bats	birds of paradise	caimans
capybara	jaguar	lemurs	orangutan	owl monkey
peccaries	piranhas	pygmy marmoset	tapirs	two-toed sloth

1. _____, piglike relations of the rhinoceros and the horse, are found in tropical America and southern Asia.

2. The _____, found in the American tropics is the only nocturnal monkey.

3. The _____, a primate of the rain forests of Southeast Asia, has a shaggy, reddish-brown coat.

4. The _____, found in the American tropics, is the world's tiniest monkey; it feeds on tree sap and gum.

5. The _____hangs upside down from tree branches and slowly travels along the branches.

6. _____ can roll into a hard bony shell when danger approaches.

7. _____, found in New Guinea and Australia, have beautiful plumage; the males have long tail feathers.

8. The _____ of South America is the world's largest rodent.

9. The _____, a nonvenemous snake of the South American tropics, constricts its prey in its coils.

10. _____, found in the American tropics, are related to alligators.

11. _____, piglike mammals of the Americas, travel in herds.

12. The _____ is a large, carnivorous, feline mammal of tropical America.

13. Many species of _____, the only flying mammals, inhabit the New World tropics.

14. _____ are carnivorous fish of the Amazon.

15. _____, which include the aye-aye and indri, are found only in Madagascar.

Grasslands

Grasslands form where there is not enough rainfall to support trees but enough to keep deserts from forming. Grasses are the primary source of food for large numbers of animals. Most grazers do not eat too close to the ground. Rather, they leave the shoots unharmed, allowing the grass to continue to grow. Of course, overgrazing by farm animals can destroy grasslands. Grasslands have local names. In North America, grassland is called the **prairie**. In Africa it is called **veld** (veldt). In southeastern Europe and Siberia it is called **steppe**. In South America it is called **pampas**. In Australia, it is called **rangeland**.

Activities

Prairie dogs are actually burrowing rodents. They live in huge colonies beneath the prairie ground. In the early 1900s, one of their largest "cities" was estimated to be home to about 400 million prairie dogs. Farmers systematically poisoned many of these animals. As one of the farmers who participated in the killings, defend your actions in a letter to the editor.

The loss of the prairie dog almost led to the extinction of the blackfooted ferret. Draw a 3-link food chain that explains the drop in ferret population.

Explain how the climate contributes to the formation of the steppe.

It is natural for grasslands to experience drought every 20 to 22 years. In the 1930s, however, drought in the prairie, or Great Plains, of the United States led to one of the worst disasters in the nation's history. Explain what is meant by the Dust Bowl and what caused it.

Savannas are flat, tropical or subtropical grasslands. The most famous is the savanna of Africa. It goes from the Sahara Desert to about the southern tip of Africa. In some areas it is mixed with thorny shrub or open woodland. Large herbivorous mammals dominate the savanna. Give three examples.

The pampas are nearly treeless grasslands of South America. Describe their location.

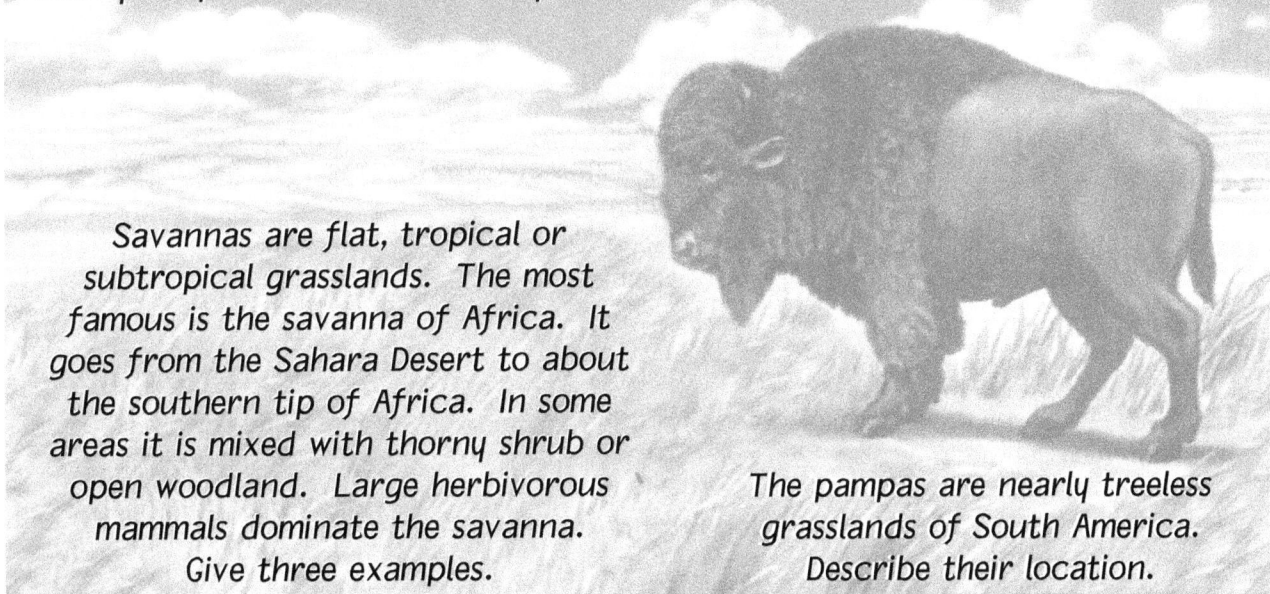

Animals of the Grasslands

Grasslands are found throughout the world. Many grassland animals in one continent have similar counterparts in other continents.

Research the animals listed below. Classify them and put each into the proper group on the next page: Leaping Herbivorous Mammals, Burrowing Mammals that Feed Underground, Burrowing Mammals that Feed above Ground, Running Flightless Birds, Running Herbivorous Mammals, and Running Carnivorous Mammals.

NORTH AMERICA
bison
coyote
ground squirrel
jack rabbit
pocket gopher
prairie dog
pronghorn

prairie dog

SOUTH AMERICA
guanaco
maned wolf
pampas cavy
pampas deer
rhea
tuco tuco
viscacha

guanaco

viscacha

ASIA
Asiatic jerboa
hamster
mole rat
pallas cat
pocket gopher
saiga
wild horse

pocket gopher

AFRICA
African ground squirrel
cheetah
golden mole
lion
ostrich
springbok
springhaas
zebra

AUSTRALIA
emu
marsupial mole
red kangaroo
Tasmanian "wolf"
wombat

red kangaroo

lion

EXTRA: When you have completed classifying the animals, create three analogies based on what you have learned.

EXAMPLE: Prairie dog : North America : : Hamster : Asia

Animals of the Grasslands

1. LEAPING HERBIVOROUS
MAMMALS

2. BURROWING MAMMALS:
FEED UNDERGROUND

3. BURROWING MAMMALS:
FEED ABOVE GROUND

4. RUNNING FLIGHTLESS
BIRDS

5. RUNNING HERBIVOROUS
MAMMALS

6. RUNNING CARNIVOROUS
MAMMALS

Life on the African Savanna

Savannas are tropical or subtropical grasslands. So many large animals live together on the African savanna that it is one of the most fascinating places on Earth. The grasslands of the eastern and western parts of the continent contain huge herds of elephant, wildebeest, giraffe, zebra, antelope, and other herbivores. Of course, lions, cheetahs, leopards, and other carnivores that prey upon these plant-eaters are also there. Most are in some way dependent upon the grass! The zebras and other grazers get their nutrients and energy directly from the grass. The lions and other predators get their nourishment from the animals that eat the grass.

Activities

Serengeti National Park is a wildlife refuge in northern Tanzania. It was established in 1951. The park has an area of about 5,700 square miles (14,763 square kilometers.) Locate Tanzania on a map.

Wildebeests, zebras, and gazelles go on long migrations in search of food. Their predators often follow. What makes these migrations necessary?

Create Who Am I? riddles of animals of the African savanna that a kindergarten child could solve.

In spite of efforts to control hunting and poaching, the rhinoceros and elephant are still being slaughtered for their ivory tusks. This has greatly reduced their populations. Write a letter to the poachers trying to convince them to change their behavior.

Animals of the Savanna

The names of these animals of the African savanna somehow got scrambled. Unscramble the letters to find out some of the diverse species to whom the savanna is home.

MAMMALS: HERBIVORES

1. T A N L E P O E _____

2. R E B Z A _____

3. S O R E C O N I H R _____

4. P H E L E T N A _____

5. A F F E R I G _____

6. D E L I W B E E T S _____

7. F B U F A L O _____

MAMMALS: CARNIVORES

8. H E E C H T A _____

9. I O L N _____

10. Y H A N E _____

11. P A R D L E O _____

12. L A K C A J _____

BIRDS

13. C O N L A F _____

14. S U B T R A D _____

15. S O T R C H I _____

16. B I L L H O N R _____

The Desert

Deserts are dry, barren places that receive less than 10 inches of rain per year. Some are covered with shifting sand. Others are barren and rocky. The little rain it does get often comes in a few brief downpours. Conditions in the desert are harsh. The lack of rain means that only a few, highly specialized plants and animals are able to live there.

Most deserts are the result of large-scale climatic patterns. They lie in a parallel belt about 25° north and 25° south of the equator. Other deserts form behind large mountain ranges. A rain shadow effect prevents the moist ocean air from moving inland. Although the term "desert" usually refers to hot deserts, there are also cold deserts. In that case the precipitation comes in the form of snow.

Activities

Draw a chart that explains why most deserts lie in a parallel belt about 25° north and south of the equator. Include a paragraph explaining your chart.

The Sahara Desert in northern Africa is the largest in the world. It spans the continent from the Atlantic Ocean to the Red Sea and extends northward from the Niger River and Lake Chad to the Atlas Mountains and the Mediterranean Sea.

Cite an example of a desert that was formed because of the rain shadow effect of a mountain range.

An acrostic is a poem or series of lines in which the first letter of each line forms a name, a word, or a message. Research the Sahara Desert. Create an acrostic about it.

Choose three plants or animals that are adapted for desert life. Explain the adaptations. Illustrate your work.

Aquatic Ecosystems

Match the aquatic ecosystems on the left with their descriptions on the right.

____ 1. Coral reef

____ 2. Estuary

____ 3. Lake

____ 4. Ocean

____ 5. Pond

____ 6. River

____ 7. Seashore

____ 8. Wetland(s)

A. The entire body of salt water covering 72 percent of Earth's surface.

B. A large, flowing body of water which empties into the ocean or other body of water.

C. Marine ridge or mound built up by coral polyps; found in shallow, sunlit waters.

D. A still body of water smaller than a lake.

E. Ground lying between the high-water and the low-water mark.

F. A lowland area, such as a marsh or swamp, which is saturated with moisture.

G. Where the river meets the sea; fresh water combines with salt water.

H. A large, inland body of fresh or salt water.

Ocean Life Cause & Effect

Match the cause on the left with the effect on the right.

_____ 1. Seawater provides a lot of support.

A. Marine plants and animals do not have to store as much energy in the form of related chemicals as land animals do and their bodies are mainly protein.

_____ 2. Swimming through water requires much less energy than lifting each foot against the force of gravity.

B. The maximum depth at which plants can photosynthesize is 328 feet.

_____ 3. There is less variation in temperature in the sea than on land.

C. Coral reefs can only grow in shallow water, where the algae get enough light to photosynthesize.

_____ 4. Only about 1 percent of the light from the surface reaches 328 feet below.

D. Marine plants and animals do not need very strong skeletons to hold their bodies in shape.

_____ 5. Reef-building coral polyps depend upon microscopic algae, called zooxanthellae, for some of their food.

E. Ocean animals and plants do not have to to cope with extremes of heat or cold or with sudden temperature changes.

_____ 6. The clown fish hides among the tentacles of the sea anemone.

F. There is less variation in temperature in the sea than on land.

_____ 7. Plants cannot grow in the Twilight Zone of the ocean, which begins at about 328 feet below the surface.

G. Fish that are afraid of being stung by the anemone also leave the clown fish alone.

_____ 8. Water heats up and cools down more slowly than air.

H. Animals that inhabit this zone must eat other animals or travel to the top level, called the Photosynthetic Zone.

Activities

There are three distinct shifts—three definite periods of activity—around a coral reef. About one-third to two-thirds of reef fishes are diurnal. They include very colorful fishes, such as butterfly fish, clownfish, and angelfish. About one-tenth are crepuscular. These include many of the large predators, such as barracuda, reef sharks, jacks, and groupers. About one-fourth to one-third of the reef fishes are nocturnal. Many have well developed senses of smell, taste, and touch. Others, such as bigeye scads, have huge eyes.

Define *diurnal, crepuscular, and nocturnal.*

Guess why many of the large predators are crepuscular.

Draw a picture representing a five-link food chain that begins with phytoplankton.

Find out what makes the cleaner wrasse unusual. Create a cartoon in which a cleaner wrasse advertises its services to other fish of the coral reef.

Unscramble the letters to figure out the name of the complex series of reefs that stretch along about 1,250 miles of Australia's northeast coast. Astronauts tell us it can be seen from the moon and even farther out in space.

ERGTA   RABREIR   EREF

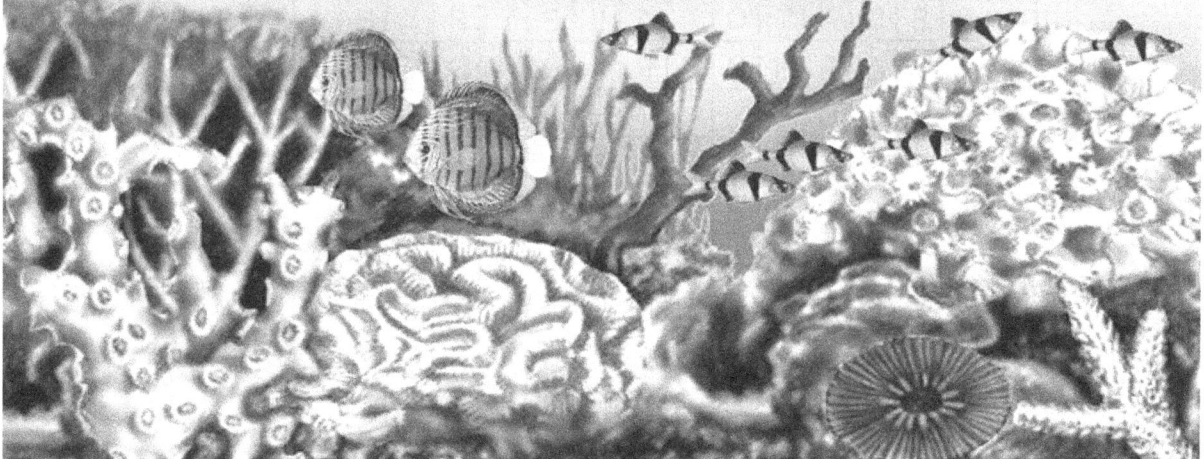

Ecology Word Search

Find the following words hidden in the puzzle and circle them. You may look in all directions: up, down, left, right, and diagonally. Circle the words.

adapt	decomposer	herbivore	savanna
biome	desert	niche	soil
carnivore	ecologist	photosynthesis	symbiosis
consumer	ecosystem	predator	tundra
coral reef	energy	producer	wetland
deciduous	habitat	rain forest	woodland

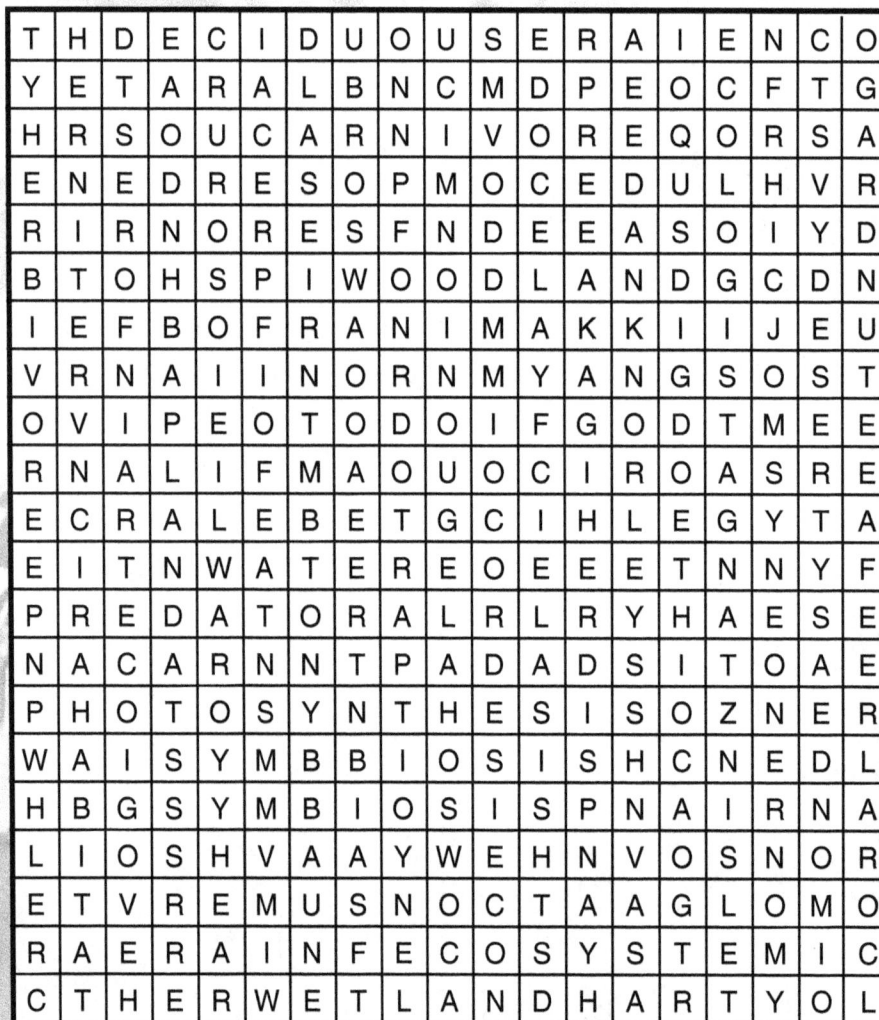

T	H	D	E	C	I	D	U	O	U	S	E	R	A	I	E	N	C	O
Y	E	T	A	R	A	L	B	N	C	M	D	P	E	O	C	F	T	G
H	R	S	O	U	C	A	R	N	I	V	O	R	E	Q	O	R	S	A
E	N	E	D	R	E	S	O	P	M	O	C	E	D	U	L	H	V	R
R	I	R	N	O	R	E	S	F	N	D	E	E	A	S	O	I	Y	D
B	T	O	H	S	P	I	W	O	O	D	L	A	N	D	G	C	D	N
I	E	F	B	O	F	R	A	N	I	M	A	K	K	I	I	J	E	U
V	R	N	A	I	I	N	O	R	N	M	Y	A	N	G	S	O	S	T
O	V	I	P	E	O	T	O	D	O	I	F	G	O	D	T	M	E	E
R	N	A	L	I	F	M	A	O	U	O	C	I	R	O	A	S	R	E
E	C	R	A	L	E	B	E	T	G	C	I	H	L	E	G	Y	T	A
E	I	T	N	W	A	T	E	R	E	O	E	E	E	T	N	N	Y	F
P	R	E	D	A	T	O	R	A	L	R	L	R	Y	H	A	E	S	E
N	A	C	A	R	N	N	T	P	A	D	A	D	S	I	T	O	A	E
P	H	O	T	O	S	Y	N	T	H	E	S	I	S	O	Z	N	E	R
W	A	I	S	Y	M	B	B	I	O	S	I	S	H	C	N	E	D	L
H	B	G	S	Y	M	B	I	O	S	I	S	P	N	A	I	R	N	A
L	I	O	S	H	V	A	A	Y	W	E	H	N	V	O	S	N	O	R
E	T	V	R	E	M	U	S	N	O	C	T	A	A	G	L	O	M	O
R	A	E	R	A	I	N	F	E	C	O	S	Y	S	T	E	M	I	C
C	T	H	E	R	W	E	T	L	A	N	D	H	A	R	T	Y	O	L

Who Ate What? Where?

A new zoo has just opened up. Unfortunately, the cards describing five exhibits have gotten mixed up. See if you can help the zookeeper get the right descriptions for the right animals by figuring out which animal ate what food and where the animal's natural habitat is. Use the clues to help you. Put an X in those boxes you eliminate. Put a √ in those that are correct.

	BANANA	FROG	ORANGE	PECCARY	TURTLE	AMAZON RAIN FOREST	AMAZON RIVER	BORNEO	MADAGASCAR	NEW GUINEA
ARAPAIMA FISH										
BIRD OF PARADISE										
FOSSA										
JAGUAR										
ORANGUTAN										

CLUES

1. One reptile, one amphibian, one mammal, and two fruits were eaten by one fish, one bird, and three mammals.

2. The animal that ate the amphibian is a catlike cousin of the civet and comes from Madagascar.

3. The fish ate a reptile and lives in a river.

4. The feathered animal ate one of the fruits; its natural habitat is New Guinea.

5. The orangutan, an herbivore, did not eat the banana.

6. The jaguar ate the mammal; its natural habitat is the place that comes before the others alphabetically.

7. The mammal who comes from Borneo did not eat the amphibian.

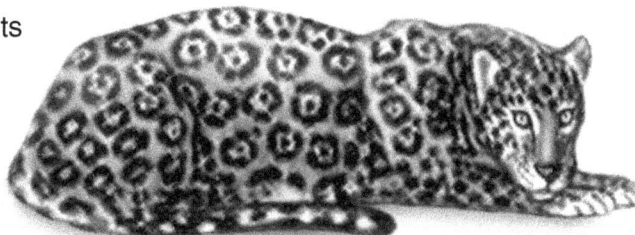

Puzzling Ecology Terms

Decode the ecology terms by solving these math clues. The fractions refer to the beginning letters of the words.

EXAMPLE: 2/3 of grown boy + 1/2 of a promise

The word clues are MAn and PLEdge. The answer is maple.

1.   3/4 of writing table + 2/5 of a mistake + 1/3 of not bottom

WORD CLUES: _____

ANSWER: _____

2.   3/4 of eating utensil + 1/3 of to consume + 1/2 of our sun

WORD CLUES: _____

ANSWER: _____

3.   3/7 of a pledge + 3/4 of a bird with webbed feet + 2/3 of to make a mistake

WORD CLUES: _____

ANSWER: _____

4.   1/2 of 5-cent coin + 1/2 of to aid

WORD CLUES: _____

ANSWER: _____

5.   1/2 of talking bird + 3/4 of largest continent + 2/3 of a beverage

WORD CLUES: _____

ANSWER: _____

6.   1/2 of bird of prey + 3/4 of to grip with teeth + 2/3 of consumed

WORD CLUES: _____

ANSWER: _____

7.   1/2 of to like better + 3/4 of day/month/year + 1/3 of red & yellow

WORD CLUES: _____

ANSWER: _____

8.   3/4 of a fish + 3/4 of to sketch

WORD CLUES: _____

ANSWER: _____

Create two more "puzzling ecology terms" of your own. Exchange with classmates to solve.

Odds 'n Ends

Create an illustrated glossary for a book on the ecosystem of your choice.

Create an annotated bibliography for that book.

Describe the African savanna from the point of view of a lion.

Create three What Am I? riddles with an ecology theme. Exchange with classmates to solve.

Describe the African savanna from the point of view of a zebra.

Plants and animals have developed a wide variety of defense mechanisms. Brainstorm and try to think of many different ones. Then make a chart showing the defense mechanisms of plants on one side and of animals on the other.

Write an acrostic about the rain forest, coral reef, or temperate deciduous forest.

"Photosynthesis" is a rather large word. See how many little words you can form by using the letters in it.* Do not use "s" to form plurals.

P-H-O-T-O-S-Y-N-T-H-E-S-I-S

*Teacher: See note in answer section before beginning this activity.

Ecology Crossword Puzzle

ACROSS

1. A large, inland body of water.
4. Any animal in a food chain.
6. Top layer of forest, formed by leaves of the tall trees.
8. Some crepuscular animals are active then.
9. When 2 different life-forms live in close association.
10. A herbivore of the tundra.
13. Hunted by a predator.
14. An ecological community of living and non-living things.
15. Active during the day.
16. Shedding leaves at the end of a growing season.
21. All food chains begin with energy from it.
22. Nocturnal animals are active then.
24. It gets passed from level to level in a food chain.
25. A succession of organisms in which energy is transferred from one organism to next.
26. Treeless area between ice cap and tree line of arctic regions.
27. Scientist who studies relationships between organisms and their environments.
28. To adjust to a situation.
29. A meat-eating animal.
30. Ecosystem which gets 100" or more of rain.

DOWN

2. Where a river meets the sea.
3. Built by 19 Down; form in sunlit, shallow waters close to the shore.
5. A flowing body of water.
7. Plant's role in a food chain.
9. A whale's habitat.
11. A community of living things of a major ecological region; usually named for dominant plant life.
12. The type of environment where an organism usually lives.
13. An organism that gets all or some of its nutrients from another with no benefit to the host.
17. A cone-bearing tree.
18. Describes 20 Down.
19. A coral, for example.
20. An arid, barren, often sandy region.
21. A tropical or semitropical grassland.
22. An organism's place and function in a community.
23. A plant-eating animal.

What's the Question?

What's the Question? is similar to the TV show "Jeopardy" in that the information is given in the form of the statement and the student responses are in the form of questions. The questions in Part 1 are worth 5 points each, and those in Part II are worth 10 points each.

Divide the class into teams of 4 or 5 students. The teacher may act as leader, or you may want to choose a student leader. The leader asks the first group a question from Part I. Whoever raises his or her hand first gets to answer. If the student answers correctly, 5 points are added to the team total. If the student answers incorrectly, 5 points are deducted. If no one wants to answer, the leader gives the correct answer and the total remains the same. If a team does not give a correct answer, the same question is then asked to the next group. If no group gets it right, the leader gives the correct answer.

When all the questions from Part I have been completed, the same rules are followed for Part II.

✵ ✵ ✵ ✵ ✵

ANSWERS TO "ECOLOGY WHAT'S THE QUESTION?"

PART I: 5 points each
1. What are producers?
2. What are herbivores?
3. What is a carnivore?
4. What is an omnivore?
5. What are decomposers?
6. What is the northern coniferous forest?
7. What is the temperate deciduous forest?
8. What is "life"?
9. What is the desert?
10. What are grasslands?
11. What are plants?
12. What is photosynthesis?
13. What are primary consumers?
14. What are secondary consumers?
15. What is a food web?
16. What is mutualism (or symbiosis)?
17. What is parasitic?
18. What is a predator?
19. What is prey?
20. What is a scavenger?
21. What is nocturnal?
22. What is diurnal?
23. What is at dusk and/or dawn?
24. What is ecology?
25. What is a niche?

PART II: 10 points each
1. What are the tropics (or equatorial region)?
2. What is the emergent level?
3. What are epiphytes?
4. What are lianas?
5. What are tree trunks?
6. What is an estuary?
7. What is a coral reef?
8. What is Australia?
9. What is plankton?
10. What is a pond?
11. What is a savanna?
12. What is a steppe?
13. What are the pampas?
14. What is North America?
15. What is top carnivore (or top consumer, etc.)?
16. What are shrubs?
17. What are herbs?
18. What is humus?
19. What are broad leaves?
20. What is detritus?
21. What is the tundra?
22. What is an ecologist?
23. What is a biome?
24. What is a lichen?
25. What is an ecosystem?

What's the Question?

Part I: 5 points

Types of Organisms	Biomes	Food Chains	Relationships	Potpourri
1. Plants are called this because they manufacture food.	6. The dominant plants in this biome are cone-bearing trees.	11. All food chains begin with these producers.	16. This term describes a close relationship between 2 organisms in which each benefits.	21. An animal active at night is this.
2. This term describes animals that eat only plants.	7. Trees that shed their leaves at the end of a growing season are dominant here.	12. It's the process by which plants use solar energy to make food.	17. One organism gets nutrients from the other without benefit to the host in this type of relationship.	22. An animal active by day is this.
3. It's another term for meat-eater.	8. The prefix "bio–" means this.	13. In a food chain, animals that eat plants are called herbivores or this.	18. An animal that hunts another is this.	23. A crepuscular animal is active during these times of day.
4. It's an animal that eats both plants and other animals.	9. The saguaro cactus is a dominant species in parts of this biome.	14. In a food chain, animals that eat the animals that eat plants are called carnivores or this.	19. An animal that is hunted for food is this.	24. It is the study of relationships between organisms and their environments.
5. Bacteria and fungi are these.	10. Grazers are dominant animals in this biome.	15. A complex of interrelated food chains in a community is called this.	20. This kind of animal feeds on dead or decaying matter.	25. It's an organism's place and function in a community.

What's the Question?

Part II: 10 points

Rain Forests	Aquatic Ecosystems	Grasslands	Temperate Forests	Pot Luck
1. Most rain forests are found in this climatic region.	6. This term refers to the area where the river meets the sea.	11. It's a tropical or subtropical grassland, such as the Serengeti Plain of Africa.	16. They are bushy plants with many woody stems.	21. Its southern boundary lies at the southern limit of the permafrost, or permanently frozen subsoil.
2. This level describes the trees that rise above the canopy.	7. It builds up slowly from the remains of polyps.	12. Grassland in Russia is called this.	17. Most wildflowers are these kinds of green plants with soft stems.	22. This kind of scientist studies living organisms and their environments.
3. These plants that grow on other plants are sometimes called "hitchhikers."	8. The Great Barrier Reef is off this continent.	13. They are the grasslands of Argentina.	18. This organic substance of partially or wholly decayed matter provides nutrients for plants.	23. It is a community of living organisms in a major ecological region.
4. These high-climbing, usually woody vines are common in the tropics.	9. These tiny, floating or weak-swimming organisms are the lowest links in the food chain.	14. Prairie refers to grasslands in this continent.	19. Most deciduous trees have this kind of leaves.	24. It is made up of a fungus and an algae living in close association.
5. Buttresses, which help brace the large trees, are really extensions of this tree part.	10. It is a still body of water smaller than a lake.	15. The lion holds this position in its food chain.	20. The bits and pieces of dead plant and animal matter on the forest floor is called this.	25. It is an ecological community of living organisms and the non-living things in their environment.

Ecology: What is it? (Page 7)
1. D   2. A   3. G   4. I   5. N   6. B   7. M   8. C   9. K   10. E   11. H   12. F   13. J   14. L

Biomes (Page 8)
Coniferous—Describing trees with needle-shaped leaves and seeds which are usually produced in a cone.

Deciduous—Describing plants that shed all their leaves in one season.

Chaparral—A dense thicket of shrubs and small trees.

Desert—A dry, barren, sometimes sandy, sometimes rocky region which receives less than 10" of rain per year.

Grassland—A stretch of rolling grass. In different parts of the world, grasslands are called savannas, prairies, pampas, and steppes.

Rain Forest—Wet and warm places that get 100" or more rain per year. **Tropical rain forests** get over 160" of rain per year and have no dry season.

Tundra—A treeless area with a permanently frozen subsoil. The tundra lies between the ice cap and the tree line of the arctic regions.

Latitude is an important factor in determining climate. Climate is the most important factor in distinguishing one biome from another.

All Kinds of Organisms (Page 9)
Without the sun's light, life on Earth could not exist. Green plants contain chlorophyll. The chlorophyll traps the sun's light energy, which plants use to form simple sugars such as glucose from carbon dioxide and water. The glucose is later converted into starches, more complex forms of food energy. This is the starting point of *all* food chains. Some animals feed on the plants. Other animals feed on the animals that feed on the plants. Still others feed on the animals that feed on the animals that feed the plants, and so on. There are seldom more than 5 links, however. Decomposers feed on both producers and consumers once these organisms are dead.

A predator is an animal that survives by hunting another for food. **Prey** is an animal hunted for food.

An autotroph is an organism that manufactures its own food. A synonym is "producer." Plants are autotrophic.

An herbivore is an animal that eats plants.
A carnivore is an animal that eats other animals.
An omnivore is an animal that eats both plants and animals.
A parasite is an organism that obtains its food from a host. It does not eat the entire body, nor does it usually kill the host.
Scavengers eat large, dead animals.
Most humans are omnivores. Vegetarians are herbivores.

What's Your Niche? (Page 10)
Diurnal animals are active during the day. **Nocturnal** animals are active at night. **Crepuscular** animals are active during the twilight hours, at dusk and/or dawn.

Some adaptations leave an organism too inflexible to adjust to new changes in their environment.

Koalas eat only eucalyptus leaves and must live where the eucalyptus trees grow. **Pandas** eat only bamboo and must live where bamboo is found.

Chains, Webs, and Pyramids (Page 13)
There aren't more links because energy is lost at each link. Animals at the top of a long chain wouldn't receive much energy.

In a pyramid of numbers for a tree the entire tree would represent the bottom, or producer level. (See picture at right.)

Partnerships, Good and Bad (Page 14)
1. Mutualism (or symbiosis)   2. Mutualism (or symbiosis)   3. Parasitism

Ecosystems (Page 15)
A steppe is a vast, semiarid, lightly wooded, grass-covered region.
A savanna is a flat, tropical or subtropical grassland.

An estuary is where the sea meets a river.

The flow of energy is accomplished through food chains.

Biomes are communities of living things. **Ecosystems** include living things and the non-living things in their environment, such as air, water, soil, and solar energy.

The Arctic Tundra (Page 16)
A possible syllogism is the following:
A. There are less than 30 days a year when the temperature on the tundra reaches 50°F.
B. Trees need at least 30 days a year when the temperature reaches 50°F.
C. Therefore, there are no trees on the tundra.

The arctic tern travels more than 22,000 miles (36,000 kilometers) from the Arctic to the Antarctic and back each year.

The polar bear and the musk ox remain on the tundra year round.

Most of the air is held close to Earth's surface by gravity; therefore, the atmosphere **high up in the mountains** is very thin and cannot hold in the heat. Nighttime temperatures drop quickly. The thin air also provides less oxygen for animals to breathe and less carbon dioxide for photosynthesis by plants. More damaging rays are also able to come through.

Northern Coniferous Forests (Page 17)
The red crossbill's beak is so specialized for picking seeds out of the cones that the cones are its only source of food.

There is a short growing season because the cold weather makes it hard for trees to draw water from the soil. Also, the sun is low in the sky and there is a limited amount of solar energy (sunlight) available for photosynthesis.

Coniferous trees have needle-shaped leaves. These leaves have a small surface, reducing the amount of water loss. The leaves are covered with a thick, waxy covering, protecting the water stored inside. Because they have their leaves when spring comes, the process of photosynthesis can begin immediately.

The taiga is the subarctic evergreen region south of tundra and north of the denser coniferous forests. There are stands of small forests of spruces, firs, and other coniferous trees. The taiga goes into the far northern regions of Scandinavia and Canada and to the Taymyr Peninsula in Siberia, which is 750 miles north of the Arctic Circle.

Temperate Deciduous Forests (Page 18)
Trees compete with each other for sunlight, moisture, and mineral nutrients. They need these for photosynthesis.

Trees help retain water and prevent water from running rapidly off the forest land during heavy rains. Their roots also temporarily hold reserves of mineral nutrients, preventing the nutrients from being washed away.

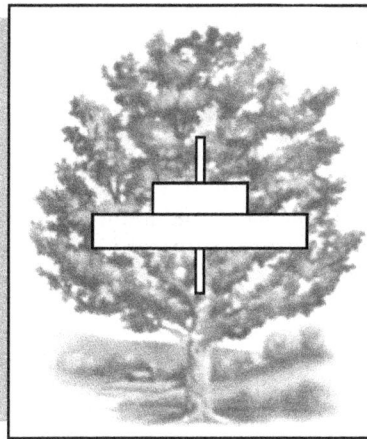

Layers of the Forest (Page 19)

Some of the tall trees of the canopy will die or be cut down. Some of the young trees of the understory will grow to replace them.

The climate is different at each level of the forest. The leaves of the trees in the canopy soak up a lot of the sunlight. They block the sun from reaching the understory and shrub levels. Even less reaches the ground. The temperature, too, is affected because of this. During the day there are higher temperatures in the canopy. At night it is about the same at all levels. Also, trees slow down the wind. They keep a lot of the rain from reaching the floor and washing away the soil. Leaves from the trees that have fallen to the floor also soak up some of the rain and keep the soil from washing away.

As the cold winter approaches, the broadleaved trees loose their leaves. Many small herbs die, surviving the winter as seeds. Other plants die but retain a supply of food reserves as a bulb or tuber. Many insects go into a state of suspended animation. Bears, bats, ground squirrels, and some other mammals hibernate. Some just become less active. Some small animals and insects die in the winter. Some animals—especially those secondary consumers whose diet is based on primary consumers that are not available—migrate.

Answers will vary, but one possible **3-link chain in the canopy** is the following: leaf → caterpillar → crested flycatcher.

Branches, leaves, petals, animal waste, and dead animals cover the ground. Decomposers, such as earthworms, fungi, and bacteria, turn this dead matter into detritus. Eventually, the bits of decayed detritus are so small that they no longer resemble the leaves or other matter; the material is then called **humus.** On the forest floor is a layer of humus between the layers of dead leaves and the minerals of the soil. Humus provides nutrients for plants and improves water retention in the soil.

Tropical Rain Forests (Page 20–21)

The tropical rain forest is called a closed, self-sustaining system because it produces its own nutrients and even its own climate. Nutrients are recycled with near perfect efficiency. As soon as a leaf, branch, or tree dies, bacteria, fungi, and other microorganisms begin to decompose it. Termites, worms, snails, ants, and other organisms help. The dead plant is decomposed and recycled into useable nutrients. The thin, shallow roots quickly absorb the nutrients. The rain forest even makes its own rain. Over half of the rain that falls stays in the canopy. The sun evaporates it back into the atmosphere. The moisture gathers in clouds and falls again as rain.

Old World tropics include the African tropics and the Indo-Malayan tropics. **New World tropics** include Central and South America.

A few large mammals of the Old World tropics are the elephant, the hippopotamus, and the gorilla.

The countries of the Amazon Rain Forest include Bolivia, Brazil, Colombia, Ecuador, French Guiana, Guyana, Peru, Surinam, and Venezuela. Most is in Brazil.

"Arboreal" means "tree-dwelling." A few examples are the red howler monkey, the owl monkey, the two-toed sloth, and the silky anteater.

The densest part of a rain forest is far from the ground, making it top heavy. To prevent strong winds from toppling them, many trees developed extensions called **buttresses.** They spread out from the bottom of the trunk and act as props, or braces. **Lianas** are thick, woody vines. They begin at the floor level and coil around the trunk and among the branches. They form a protective net around the trees and throughout the canopy, preventing the trees from falling over even in strong winds. Lianas are used to make rattan for wicker furniture.

About one-fourth of all prescription drugs come from plant species found in the rain forest. Among the diseases treated by tropical plants are malaria, Hodgkin's disease, childhood leukemia, and rheumatoid arthritis. Mr. Thomas Lovejoy of the Smithsonian Institution referred to the Amazon Rain Forest as "the world's greatest pharmaceutical laboratory."

Jungles are usually created when people cut down canopy trees. In a rain forest very little sun gets through the thick canopy to the undergrowth. Once the canopy trees are gone, the sun hits the plants on the floor level, causing these plants to grow and multiply rapidly.

Unlike most plants, **epiphytes** begin their growth in the canopy rather than on the forest floor. Some of their roots dangle over branches and absorb water from the moist air. They use their roots to anchor themselves onto the branches; however, they do not get their nutrients from the trees. Several species of orchids, mosses, lichens, bromeliads, and ferns are epiphytes.

Plants depend on animals in several ways. Birds, such as the hummingbird, carry pollen from flower to flower, helping to fertilize them. Trees depend on animals to spread their seeds; if all the seeds remained where they fell, there would be too much competition for light, space, and nutrients and they wouldn't survive. Monkeys and other animals drop pits away from the source. Seeds also go through the digestive tracts and are scattered.

Rain Forest Fact or Opinion (Page 22)
1. O 2. F 3. F 4. O 5. F 6. O 7. F 8. F 9. O 10. O 11. F 12. O 13. O 14. F

Animals of the Rain Forest (Page 24)

1. tapirs	4. pygmy marmoset	7. birds of paradise	10. caimans	13. bats
2. owl monkey	5. two-toed sloth	8. capybara	11. peccaries	14. piranhas
3. orangutan	6. armadillos	9. anaconda	12. jaguar	15. lemurs

Grasslands (Page 25)
The 400 million **prairie dogs** could have eaten as much grass as 1.5 million cows. The farmers wanted to prevent the loss of grassland.

The **3-link chain** would be as follows: grass → prairie dog → ferret. The ferrets' food supply was gone.

The Dust Bowl was caused by years of severe drought, made even worse by the farming practice of yearly plowing which broke the roots and made it easier for the soil to dry out and blow away. During the Dust Bowl, clouds of dust blew from the prairie to areas as far away as the Atlantic Ocean. In the region itself, dirt covered just about everything: fields, machinery, barns, and houses. It even filled the houses, covering flooring, furniture, clothing, and food. People and animals died of respiratory diseases.

On the African savanna zebras, gazelles, and wildebeests roam in huge herds. Others are giraffes, buffaloes, rhinoceroses, and elephants.

The pampas are located mainly in central Argentina and Uruguay, east of the Andes and south of the Amazon Rain Forest.

Animals of the Grassland (Page 26–27)
The following animals should be listed under **Leaping Herbivorous Mammals:** jack rabbit, Asiatic jerboa, African ground squirrel, springhaas, and red kangaroo. The following animals should be listed under **Burrowing Mammals that Feed Underground:** ground squirrel, prairie dog, pampas cavy, viscacha, hamster, and wombat. The following animals should be listed under **Burrowing Mammals that Feed above Ground:** pocket gopher, tuco tuco, "mole rat," golden mole, and marsupial mole. The following animals should be listed under **Running Flightless Birds:** rhea, ostrich, and emu. The following animals should be listed under **Running Herbivorous Mammals:** bison, pronghorn, guanaco, pampas deer, saiga, wild horse, springbok, and zebra. The following animals should be listed under **Running Carnivorous Mammals:** coyote, maned wolf, pallas cat, cheetah, lion, and Tasmanian wolf.

Life on the African Savanna (Page 28)
The **migrations** are necessary because Africa has two seasons, a rainy one and a dry one. During the dry season food becomes scarce.

Animals of the African Savanna (Page 29)

1. antelope	3. rhinoceros	5. giraffe	7. buffalo	9. lion	11. leopard	13. falcon	15. ostrich
2. zebra	4. elephant	6. wildebeest	8. cheetah	10. hyena	12. jackal	14. bustard	16. hornbill

The Desert (Page 30)
Most deserts lie in a belt 25° north and south of the equator. The sun is always overhead somewhere in the tropics. It is over the equator the longest, warming the land. The land, in turn, heats the air above it. The hot air rises, carrying moisture from the sea and land. As it rises, it cools. When it cools, it sheds its water, creating rain forests. By the time it gets to the Tropic of Cancer and

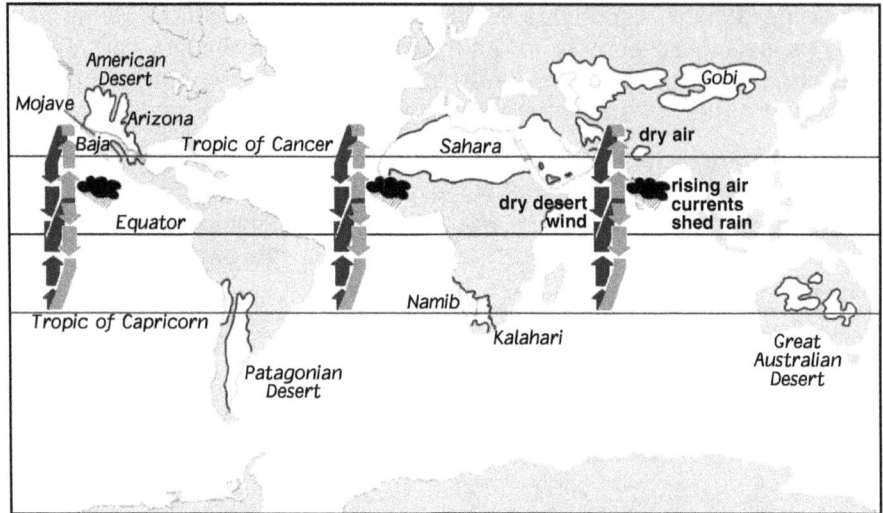

Tropic of Capricorn, it has lost its water. The hot air is drawn back towards the equator, creating hot, dry winds over the desert area. The winds draw moisture from the air, making it even drier.

Possible examples of **deserts that were formed because of rain shadow effects** include the following: Mojave Desert of Nevada and eastern California (Sierra Nevadas); Gobi Desert of central China (Himalayas); and the deserts of eastern and central Australia (Great Dividing Range).

Aquatic Ecosystems (Page 31)
1. C   2. G   3. H   4. A   5. D   6. B   7. E   8. F

Ocean Life Cause & Effect (Page 32)
1. D   2. A   3. E   4. B   5. C   6. G   7. H   8. F

Ocean Life Activities (Page 32)
Diurnal means active by day; *crepuscular* means active at dawn and/or dusk; and *nocturnal* means active by night.

The large predators, although active at all times, are especially active at dawn and dusk because it is easier to surprise their prey in the confusion of the changing of the shifts.

A possible chain would be as follows: phytoplankton → zooplankton → small fish → larger fish → sea mammals.

The cleaner wrasse cleans bits of food from the teeth and gills of other fish. It also removes lice from their skin. Sometimes the fish line up, waiting their turn. The cleaner wrasse benefits by eating what it removes.

The Great Barrier Reef is off the coast of Australia.

Ecology Word Search (Page 34)

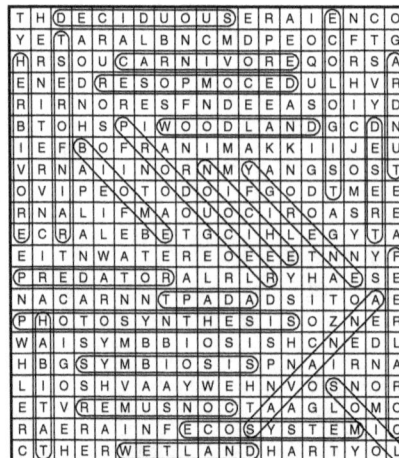

Who Ate What? Where? (Page 35)

The **arapaima fish** ate the turtle; its natural habitat is the Amazon River.

The **bird of paradise** ate the banana; its natural habitat is New Guinea.

The **fossa** ate the frog; its natural habitat is Madagascar.

The **jaguar** ate the peccary; its natural habitat is the Amazon Rain Forest.

The **orangutan** ate the orange; its natural habitat is Borneo.

	BANANA	FROG	ORANGE	PECCARY	TURTLE	AMAZON RAIN FOREST	AMAZON RIVER	BORNEO	MADAGASCAR	NEW GUINEA
ARAPAIMA FISH	x	x	x	x	✔	x	✔	x	x	x
BIRD OF PARADISE	✔	x	x	x	x	x	x	x	x	✔
FOSSA	x	✔	x	x	x	x	x	x	✔	x
JAGUAR	x	x	x	✔	x	✔	x	x	x	x
ORANGUTAN	x	x	✔	x	x	x	x	✔	x	x

Puzzling Ecology Terms (Page 36)

1. desk, error, and top = desert
2. fork, eat, and star = forest
3. promise, duck, and err = producer
4. nickel and help = niche
5. parrot, Asia, and tea = parasite
6. hawk, bite, and ate = habitat
7. prefer, date, and orange = predator
8. tuna and draw = tundra

Odds 'n Ends (Page 37)

Possible answers for **plants' defense mechanisms** include thorns, prickles, leathery leaves, spines, growing among bushier or taller plants, patterned leaves (camouflage), unpleasant or poisonous chemicals, and unpleasant-tasting oils. Possible answers for **animals' defense mechanisms** include speed (running or flying), herds or flocks, antlers or horns (often just for display), well-developed senses of sight or hearing, camouflage (shape, color, stripes, spots, and seasonal changes), trickery, foul-smelling secretions, bad taste, mimicry (of bad-tasting species, etc.), and bright colors to warn of poison.

Over 100 smaller words can be formed from the letters in photosynthesis. I found about 140; I'm sure there are others. Unfortunately, **among the words are a few which are inappropriate for the classroom** (shit, snot, and snotty). Use your judgment as to how to handle this depending upon the age and maturity of your class. One possibility would be to instruct the students not to include slang. The words I found are the following: eon, hen, hint, hip, his, hiss, hit, hoe, hone, honey, hoop, hoot, hop, hope, hose, host, hot, hype, into, ion, net, nit, nosh, noose, nose, not, note, one, open, opine, pen, pent, peon, peony, pest, pet, phone, phoney, phony, photo, pie, pine, pint, pit, pith, pithy, pity, poet, point, pointy, poise, pony, pose, post, pot, potty, sent, set, she, shin, shine, shiny, ship, shoe, shone, shoot, shop, shot, shote, shy, sin, sine, sip, siphon, sis, sissy, sit, site, snip, snipe, snit, snoop, snoot, snooty, son, soon, soot, soy, spent, spine, spit, spite, spoon, spot, spotty, spy, stein, step, stet, stint, stone, stoop, stop, sty, synthesis, syphon, ten, tent, test, the, then, theses, thesis, thin, thine, this, those, tie, tin, tint, tiny, tip, tiptoe, toe, ton, tone, too, toot, tooth, toothy, top, toss, tote, type, typhoon, yen, yes, yet, and yon.

Ecology Crossword Puzzle (Page 38)

What's the Question? (Pages 39–41)

Answers are given on page 39.

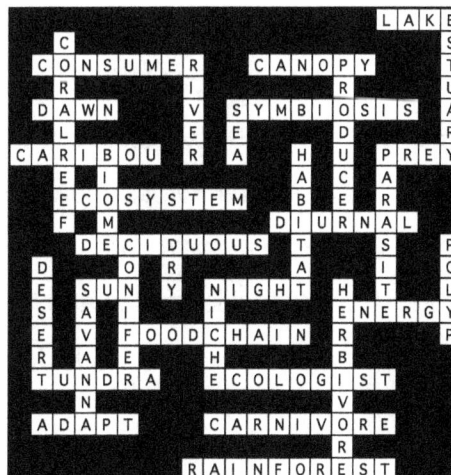

Bibliography

Battan, Mary. *The Tropical Forest: Ants, Animals, and Plants.* New York: Thomas Y. Crowell Company, 1973.

Cerullo, Mary M. Coral Reef, *A City that Never Sleeps.* New York: Penguin Books, Inc., 1996.

Farb, Peter and the editors of Time-Life Books. *Ecology.* New York: Time-Life Books, 1970.

Hoke, John. *Ecology.* New York: Franklin Watts, 1977.

Lauber, Patricia. *Who Eats What?: Food Chains and Food Webs.* New York: HarperCollins Publishers, 1995.

Milkins, Colin S. *Discovering Pond Life.* New York: The Bookwright Press, 1990.

Naden, Corinne J. *Woodlands around the World.* New York: Franklin Watts, Inc., 1973.

Nightwatch: Nightlife in the Tropical Rain Forest. Pleasantville, New York: Reader's Digest Children's Books, 1999.

Oates, John. The Living Earth, Volume 16: *Web of Life.* Danbury: The Danbury Press, 1975.

Pringle, Laurence. *Chains, Webs, and Pyramids.* New York: Thomas Y. Crowell Company, 1975.

_____. *Coral Reefs, Earth's Undersea Treasures.* New York: Thomas Y. Crowell Junior Books, 1988.

_____. *Estuaries, Where Rivers Meet the Sea.* New York: Thomas Y. Crowell Junior Books, 1973.

_____. *The Gentle Desert.* New York: Thomas Y. Crowell Junior Books, 1977.

_____. *Into the Woods, Exploring the Forest Ecosystem.* New York: Thomas Y. Crowell Junior Books, 1988.

Savan, Beth. *Earthwatch, Earthcycles and Ecosystems.* New York: Addison-Wesley Publishing Company, Inc., 1991.

Scott, Michael. *The Young Oxford Book of Ecology.* New York: Oxford University Press, 1995.

Sayre, April Pulley. *Desert.* New York: Henry Holt and Company, Inc., 1994.

_____. *Grasslands.* New York: Henry Holt and Company, Inc., 1994.

_____. *Taiga.* New York: Henry Holt and Company, Inc., 1994.

_____. *Temperate Deciduous Forest.* New York: Henry Holt and Company, Inc., 1994.

_____. *Tropical Rain Forest.* New York: Henry Holt and Company, Inc., 1994.

_____. *Tundra.* New York: Henry Holt and Company, Inc., 1994.

Warburton, Lois. *Rainforests.* San Diego, CA: Lucent Books, Inc., 1991.